Irregularities, Frauds and The Necessity of Technical Auditing in

Construction Industry

Irregularities, Frauds and The Necessity
of Technical Auditing in

Construction Industry

A.L.M.Ameer MRICS, AAIQS, ACIArb.

Chartered Quantity Surveyor & Arbitrator

authorHOUSE®

AuthorHouse™ UK Ltd.
1663 Liberty Drive
Bloomington, IN 47403 USA
www.authorhouse.co.uk
Phone: 0800.197.4150

Published by AuthorHouse   07/10/2013

ISBN: 978-1-4817-9975-1 (sc)
ISBN: 978-1-4817-9974-4 (hc)
ISBN: 978-1-4817-9976-8 (e)

Contents

1

CONSTRUCTION IRREGULARITIES

Construction Industry is huge and is growing practically in every country in the world. Professionals who are involved in the construction industry have a professional responsibility to give the necessary confidence the industry needs.

Like in any other works, there are lot of irregularities taking place in construction. These irregularities do cost money and time to the owners, be it an individual Client, who is building up a house to himself to that of a multi-million dollar investor doing a mega project. The only difference is the amounts which are lost in construction. There is no definition for irregularities, but what is not regular is considered to be irregular. There is a thin line between irregularities and frauds. A professional has to do a job, for which he is paid for. If you ask a layman about his construction of a small house, he will give you a lecture on the problems, the losses and other difficulties he had. The case is same, if it is a major project. The only difference is the amount of loss which is small for a small Contract, and much bigger running into millions of Dollars for a major project.

Some recent findings are given below to indicate the level of irregularities in the construction field that are happening worldwide on a regular basis.

China has investigated and handled more than 21,000 cases of irregularities in construction projects in 2009.Among the 21,766 cases, 3,305 or 15.2%, is occurring during the bidding process. The result indicated corruption in this sector includes interfering with the bidding process, insider trading and cheating. The report urged all relevant departments to carry out further investigation to crack down on irregularities in accordance with the relevant laws and cadres who misuse authority and interfere with the bidding process should be severely dealt with.

Another report by a central office in charge of the cleanup said more than 15,000 Chinese officials have been punished for graft or dereliction of duty over the two years (2009-2011) during a nationwide check against irregularities in the construction sectors, The situation to say the least is frightening as according to certain reports, China will be the leading construction market by 2018.By 2020, the construction market in China is worth around US $ 2.5 trillion which is about 19% of the global construction output.

A private newspaper in Bhutan-the Bhutanese says at least 50%of Nu.435 million that was spent on the construction of the three domestic airports in the country have been lost in illegal over—payment or have been wasted through avoidable expenditure.

A report indicated that the European Union has frozen Euro 890 million in aids earmarked for building Polish roads because of suspected frauds by contractors, according to the commission appointed for same. The suspension, which is likely to be only temporary, comes at an awkward time for the Polish government, which is already dealing with a sharp economic slowdown and thousands of layoffs in the construction industry.

Polish officials said they were doing everything they could to root out the suspected fraud.

The commission said it acted after Poland launched a criminal investigation into allegations that there was a price-fixing cartel involving companies bidding for road building contracts.

African countries are full of construction irregularities. Even developed countries such as the United Kingdom and the United States are not immune to irregularities in construction.

One fails to understand how such things are possible in government funded construction projects, where there is an Auditing Department with wide powers for the Auditor General. It should be also noted that, there are double auditing (internal & external) for government funded projects. Sad to say, it is happening and continues to happen. The author feels that Technical Auditing with proper guide lines and by competent Engineer Technical Auditors may solve this problem to a great extent.

The Construction professionals should be aware of their job regarding all aspects of construction. They should treat every case separately as everyone is unique and important to the Employer, not just another case in their long career. There should be a continuous learning process for the Engineers. We hear cases, where construction started without proper understanding of the scope leading to changes, which cost time and money, The Government Service Authority regulations shall be the reasons for delay and eventual late completion. All construction professionals to have check lists, a basic of which is given under sub heading 1.5 Check List, to make the project reasonably good and finished within time and budget.

It is a necessity for the contractors to remember the importance of documentation such as contracts, pricing, backups, correspondence etc. as at the end of the day contractors shall be in a position to lose lot of money as they are unable to claim owing to the lack of proper documentations, for the work, which were completed

The construction irregularities often remind me, the proverb – It is play for the cat where as it is death for the Rat as cat always play with the rat before killing it. These irregularities do cost dearly for the Employers. Just, I'll quote a few simple cases to illustrate the point.

1.1.1 Case One: Project over Run—Time

An Engineering Consultant supervised project of 7 million with a Contract period of 10 months took another 14 months to complete. The Consultant Fees for supervision amounted to 1,500,000/=.The Employer terminated the Contract of the Consultant one month after Taking over Certificate. The Employer continued to give work in patches during Defects Liability Period. The Contractor has submitted the Draft Final Account, which included all final measurements and additional works. The Employer is in a difficult position as he is unable to finalize the account of the project, as the Consultants have been demobilized and most of the Contractor staff who was involved in this particular job has already left.

1.1.2 Case Two—Project over run both Cost & Time

The common irregularity in most if not all construction projects are the project over run both in terms of time and money. Often Employers are reluctant to pay the additional money but much more accommodating to give extension of time without associated costs. The Employer actually loses in the event of delay as he is unable to get the return on the investment and that is the reason for the contractual provision of liquidated damages for delay on the part of the Contractor. While European courts are reluctant to enforce Penalty for delay as it is some form of deterent, but courts always enforced the liquidated damages if the same is found to be reasonable.

There are various factors for delay. The delay can be caused by the Contractor for lack of resources, by Engineer owing to correction of design, changes, and late approvals or by Employer by changes or can be other reasons such as weather, strikes etc.

The Contractor to a great extent (as there is no necessity to pay Liquidated damages) and the Engineer to a small extent (as corrections to design or late approvals will not be highlighted) happy to give Extension of time without associated costs.

Project cost/time overruns is a global problem applies equally to developing countries to highly developed countries with sophisticated delivery systems. A global survey of the construction sector spanning twenty countries and five continents found interesting reading, as follows:

- Substantial cost escalation on construction and infrastructure project is the rule rather than the exception.

- The average cost escalations of 45% for rail projects, 34% for tunnels & bridges and 20% for roads.

- 90% of the construction projects had under estimated the costs and that cost overruns of 50-100% were common.

- In India approximately half of all road projects have cost over runs greater than 25% and time extensions exceeding 50%

A major report by the World Bank in 2010 has recognized the global problems in construction as the industry is valued at US $ 1.70 trillion worldwide with a significant proportion involving publicly financed projects. Some examples of recent global major project cost overruns are given below:

- Boston's Central Artery/Tunnel project—275%over budget amounting to US $ 11billion

- The Channel Tunnel between UK & France 80% over budget for construction and 140% over financing

- The Pentagon spy satellite programme—US $ 4 billion cost overrun

- International Space Station—US $ 5 billion overrun

There are so many other projects with the similar situations and the list is endless generally resulting in considerable wastes.

Unfortunately these cost overruns are not new. The Suez Canal costed 20 times the original budget and the Sydney Opera House 15 times.

As the projects become bigger, the problems with cost overruns and benefit shortfalls also grow bigger. Some mega projects are becoming so large in relation to national economies that cost overruns and benefit shortfalls from even a single project may destabilize the finances of an entire country or region. The example of which was noticed when the billion-dollar cost overrun on the 2004 Athens Olympics affected the credit rating of Greece, some shocks of which are still being felt. A similar situation occurred when the benefit short falls hit Honk Kong's new 20 Billion Cheek Lap Kok airport after it opened in 1998.

There are three main causes as follows:

Technical

Imperfect forecasting techniques, inadequate data, honest mistakes, inherent problems in predicting the future, lack of necessary know how or experience on the part of the forecasters. This may be reduced or eliminated by the development of better forecasting models, much better and reliable data along with more experienced and talented forecasters.

Psychological

This what psychologists call the planning fallacy and optimism bias. In the grip of the planning fallacy, planners and project promoters make decisions based on delusional optimism rather than on a rational weighting of gains, losses, and probabilities. They over estimate benefits and under estimate the costs. They involuntarily spin scenarios of success and overlook the potential mistakes and miscalculations. As a result, planners and promoters pursue initiatives that are unlikely to come in on budget or on time, or to ever deliver the expected returns. Over optimism can be traced to cognitive biases, that is, errors in the way the mind processes information. These biases are thought to be ubiquitous, but their effects can be tempered by simple reality checks, thus reducing the odds that people and organizations will rush blindly into unprofitable investments of money and time.

Political-Economic

Planners and promoters as deliberately and strategically over estimating benefits and under estimating costs when forecasting the outcome of projects. They purposely spin scenarios of success on paper and gloss over potential or anticipated failiures.This is done in order to increase the likelihood that, it is their project, and not the competitions that gain approval and funding

Large infrastructure projects are loan financed and have long construction periods, they are particularly sensitive to delays and cost overruns as these result in increased debt, increased interest payments, and much longer payback periods.

Large cost overruns and benefit shortfalls tend to destabilize policy, planning, implementation and operations.After several cost overruns in the phase of the Sydney Opera House, the Parliament of New South Wales decided that every further ten percent increase in the budget would need their approval. Then the Opera House received lot of kicks and became a hot topic with unfriendly debate in parliament when approvals are required for additional funding. Ultimately produced an Opera House not suited for Opera. Many other projects suffered the same fate.

The UK study shows that strong interests and strong incentives exist at the project approval stage to present projects as favorable as possible, that is with benefits highlighted and cost and risk dehighlighted. Local authorities, developers, land owners, labour unions, politicians, officials and consultants all stand to benefit from a project that looks favorable on paper and they have little incentive to actively avoid bias in estimates of benefits, costs and risks.

Finally, the UK and US studies arrive at results that are basically similar. Both studies indicate well that existing data on cost under estimation and benefit over estimation. In situations with high political and organizational pressure, there is always this tendency of lowering of costs and increasing the benefits is caused by non-intentional technical error or optimism bias. Both studies support the view that in such cases promoters and forecasters intentionally used the formula—Under estimated costs + Over estimated benefits = Project approval, in order to secure the approval and funding for their projects.

It is clear that planning and implementation of large infrastructure projects need reform; less deception and more honesty are required in estimation of costs and benefits if better projects are to be implemented. It is also to be understood that costs and benefits are not the only basis for deciding whether to build large infrastructure projects. Therefore forms of rationality other than economic rationality, a balanced in the broader frame of public decision making process is required for large infrastructure projects.

Some positive actions are taking place around the world regarding mega projects, particularly in democratic countries practicing good governance. In 2003 the Treasury of the United Kingdom required for the first time, that all ministries develop and implement procedures for large public projects that will curb what it calls-with true British civility-"optimism bias. Funding will be unavailable for projects that do not take into account this bias, and methods have been developed for how to do this. In the Netherlands in 2004, the Parliament Committee on Infrastructure Projects for the first time conducted extensive public hearings to identify measures that will limit the failures of such projects. Various researches were carried out in Malaysia regarding the ethical issues that occur during pre-stage of mega project procurement. These can be considered as good signs for the future.

Projects costs & time over runs of massive projects can be tolerated as these are generally one—off. After all, we are not going to build Suez Canal or Sydney Opera House every year. What cannot be tolerated to an ordinary man in the street, let alone a Construction Cost professional is the repeated financial failures in ordinary constructions such as a school building or a shopping mall, that too in the government funded construction projects, where there is an Auditor General with wide powers and a set of Auditors attached to the Department, who are supposed to perform double (internal & external) auditing. The Author is only interested in having a Technical Auditing System for normal not so complicated regular construction in the Government sector. The Author also feels that there is definite room for improvements in the systems given in this book and the same to be further developed by Professional Institutions, Universities and big construction projects finance lending Institutions to suit their requirements as for example the Technical Auditing programme for a high way project need not be exactly the same to that of a housing complex.

1.1.3 Case Three—Not getting "Value for Money"

Any villa or house construction is required to have an Engineering Consultants approval to proceed. So some Engineering Consultants have standard villas or houses, (say two bedroom, 3 bedrooms, 4 bedrooms etc.)which they show to their Clients. The Clients often make some changes and with the amended drawings, the Consultant gets a few quotations and awards the same to the lowest and gets their fees. The Consultant does not prepare proper tender documents for example with a standard Bill of Quantities to give a uniform basis for tender, and to make adjustments in the event of additions, omissions during implementation. Do not do research to see the applicability of the Base Programme (if there is one) in relation to the available resources with the contractor, materials availability in the market etc. in order to get value for money. The end result is the Employer suffers with delays and other financial problems on a regular basis. Projects are never finished on time and forced into a situation, where they are unable to terminate and bring another Contractor to do the work.

1.1.4   Case Four; Consultant's fault—Employer escaped because of the fault of the Contractor

In a major project, the Phase 1 was for Piling. This was a project with duration of 40 days. During implementation of the project, the Electrical Sub-Station at the site was shifted as the substation capacity required upgrading owing to the new building. This resulted in hold up of several weeks as new higher capacity cables to be layed.The Contract resulted in a "Time at Large situation. Fortunately, the Contractor didn't have suitable personal such as Claim specialist to make a claim for idling equipment. There was a Claim much later after completion, which was rejected owing to lack of Notice.

1.1.5   Case Five—Smart play by Contractor

The Electrical design was changed by the Service Authorities during implementation, which resulted in the shifting of Chillers. The end result is 4C x300 mm Armored cable increased from 100 m at tender to about a Kilometer. It was found later, that the Electrical cables were over priced at tender around 300% more than the market prices. We had lot of difficulties in finalizing the Contractor entitlement for this item of work. We do not know, whether it was just a coincidence or the Contractor over priced the Electrical items in anticipation of increase during implementation.

All these are not frauds, but irregularities, which give openings to smart Contractors to play. This kind of play by the Contractors shall not come to light in the Financial Auditing that are practiced in most countries.Therefore,it is very important for the Construction professionals to be aware of irregularities and check everything before the commencement of work.

1.1.6   Case Six: Consultant's fault resulting in loss to Employer

In one Lump sum project, the original Contract price was 28 million and finally ended up as 40.50 million. The reasons are given below:

1. The Consultant missed lot of requirements for the project, which were later added as Variations with new prices for new items.

2. Some materials specified in the tender documents by the Consultant were no longer produced and not available in the market. This required variation orders and new materials and new prices.

3. Delay by the Consultant and Engineer requiring Extension of Time and Associated Costs.

Let us now analyze the losses to the Client in detail.

Projects are contracted on the basis of Lump sum Contracts to have certainty of time and cost. The Consultant is paid a fee to do his job in a professional manner in order for the Client to get the best value for money. In this instance the Consultant has failed in his professional duty of exercising the necessary skill and care pertaining to their profession.

The errors in the documents are very major such as missing the necessary requirements for the functioning of the project, specifying materials that are no longer produced(may be copy & paste from some very old document).These subsequent changes do cost time owing to approvals.

The owner loses in many ways such as

1. Not getting the anticipated return on the investment, unable to collect delay penalties from the contractor. In addition the Client has to give extension of time, which is further loss as no return is possible and further additional costs to the Client.

2. Generally all variations are priced higher and it is common knowledge in the construction circles that all variations are inflated at least by 10% when compared to the tender prices.

3. The Client is placed in a very uncomfortable position financially.In this instance, the Contractor was lucky that the Client was able to absorb the additional financial burden. In some cases, this kind of increase (45%) shall make it unbearable to the Client and may even abond the project half way incurring losses to all parties.

4. There is a possibility for the Contractor to play by pricing the items low which are Likely to be removed and higher where there can be additions.

1.1.7. Case Seven: Consultant Fault resulting in huge loss

Mega Build, Own, Operate, Transfer (BOOT) projects are very popular now. Some errors in the feasibility studies (as the study has to be period of 50 years) can result in a project being not economically viable and end up as a white elephant. The success rate of these projects stand around 20%, therefore improvements in the analysis of such projects are required as huge sums of money are involved in such projects.

1.1.8. Case Eight: Professional mistakes resulting in big loss to Clients

Some Consultants do not study carefully the Client's requirements and do give unrealistic budget values. It is the responsibility of the Consultant to have data base or other details to enable them to prepare approximate estimates. These to be checked and updated on a regular basis. The non-availability of such proper estimating tool coupled with non-proper analysis of the Clients requirements result in abortive work and a huge loss to Clients. An example is given below:

A Client wanted to build a 3 Star Hotel and approached a consultant. This Consultant has constructed similar hotels and had a data base which dealt with the cost per room. Unfortunately, the consultant has not done any hotels for the last five years and also did not update their data base. The Client was given a budget price based on the number of Rooms, which happened to be 200.

The Client was happy as he had sufficient budget to do the project. An agreement was signed between the Consultant and Client to do the design & Tender documentation and the Consultant Fees was agreed as 2.50% of the Final Tender value.

The project was tendered among the pre-qualified contractors. The lowest Tender was 60 % above the budget. The Client was unable to do the project as the amount is beyond his means. The consultant got away with his fees. The other interesting thing in this whole scenario was that the Consultant wanted his fees to be calculated on the basis of the lowest tender, which is 60% more than the budget given by him. The Client while reluctant to pay anything, as he was misguided in the first place, finally agreed to consider payment based on the original budget.

It is very important for the Technical Auditor to audit projects, which are abounded (after design) owing to budget constrains in government funded projects.

1.1.9 Case Eight: Possibility of Fraud by Contractor

Most small scale construction are lump sum projects.However, the Substructure is subject to re-measure as these cannot be estimated at the time of tender. Most Engineering Consultants do not have Land surveyors and trust the Contractor Surveyor for the levels both existing and final. There are possibilities for irregularities and it is very difficult or impossible to check.

The case is very serious in projects involving heavy earth works, such as Parks, Roads

1.1.10 Professional Indemnity Insurance

There is a necessity for the Engineering Professionals to have Professional Indemnity Insurance to compensate their financial losses to Clients. Clients should insist regarding the Professional Indemnity Insurance of their Engineers. The cover should include for Professional negligence. It is much easier for the Client to claim from the Insurance Company, rather than the lengthily legal process in the event of loss owing to the mistakes of the Engineer.

Case One

The Client appointed the Engineer for Design and a Cost Engineer to look after the costing side of the project. It was decided to treat this as a Design & Build Lump sum project meaning the Drawings and specification take precedence. A Bill of Quantities was prepared by the Cost Engineer in order to tender the project among pre-qualified Contractors to obtain a competitive price. While preparing the Bill of Quantities, the Cost Consultant has made an error—the steel requirement was given as 1,500 tonnne instead of 150 tonne.The Contractor, who was the lowest was awarded the project. The Client later found that the awarded Contractor has priced the steel for 1,500 tonnne (as given in the BOQ) instead of the required 150 tonne and claimed the additional money from the Contractor as it is a mistake. The Contractor refused stating that this is a Lump sum Contract, where Drawings and Specifications take precedence and there were so many other mistakes in the Bills of Quantities and adjustment for one item is not acceptable. The

Client then claimed the same amount from the Cost Engineer as this is a professional negligence on his part.

We have to take the following into consideration:

The Courts generally take the view that Professionals should up hold the necessary skill and care pertaing to their profession.

To claim damage, there should be a material or monetary loss.

Some Courts held that BOQ Rates should take precedence and used as base for Variations as it was proved that these BOQ Rates are not there as numbers .but undergone a professional analysis before the award. Such Tender Analysis are discussed with the Client in the context of best possible value for money spent and it is the responsibility of the professional to bring any such discrepancies and suitable actions taken before the award.

The Authors comments are as follows:

1. There is a loss to the Client owing to the mistake by the Cost Engineer and the Client to be compensated. If there is a Professional Indemnity to cover the professional negligence, then Client gets his money from the Insurance.

2. The Cost Professional has failed not once but twice, first making the error in the BOQ, the second by not picking up the same in the subsequent Tender Analysis. It is the accepted practice as "Value for Money "to do a Tender Analysis before the Award, and particularly as this is a Design & Build Lump sum Project, where it is not re-measured upon completion and the accepted price is the final value of the project.

3. The Author's assumption is based on the following:

    3.1. In this particular tender, there is every possibility for at least in one, if not many tenders received that there will be much and apparent difference in the amounts between the tenderers for this Item of vwork, as generally and particularly in Lump sum Contracts, the Contractors usually check the quantities approximately, (not necessarily in detail).This kind of discrepancy shall definitely be apparent. The Tenderers either bring such kind of mistakes to the Engineer for correction or use such mistake to their advantage. This gives the explanation for the wide difference in the amounts.

    3.2. The contractor cannot be penalized for errors in the contract documents as it was not prepared by him. It is an accepted practice in legal cases that the party who has made the mistake (in this case the BOQ preparation) cannot benefit from such mistakes.

    3.3. The question of unjust enrichment cannot apply as the amount involved is small in comparison to the tender value. The onus of proof which is the responsibility of the Client in this case is difficult as the amount involved is around 2% of the Contract price.

    3.4. All Professional Indemnity Insurance shall generally cover for the losses suffered in the event of mistakes done by the Professionals, which results in monetary loss to Clients.

Case Two—Status of Foul Drainage

The construction of a major project can take upto two years. Lot of things can happen in relation to the Foul drainage system during that period. A major housing project consisting of 500 units was designed and supervised by Engineering Consultants. The Design portion included Preliminary study, Preliminary and Final design. The scope of the Preliminary study included among others the Foul drainage system of the houses. The project was awarded via a proper procurement route as Re—measure contract. It was found later by the Technical auditors that the Employer incurred substantial loss owing to abortive works. The detailed study highlighted the cause of the abortive work as the project was implemented with soakage pit and septic tank for each villa. Subsequently large no of houses Foul drainage was connected to the main network passing around that area. The abortive work was the work carried on such tanks as per programmed and later backfilled with existing or imported materials.

The Author feels that the Employer can claim from the Professional Indemnity Insurance as it is an avoidable loss owing to the mistake of the Engineer as his scope for design includes the necessary studies regarding the Foul drainage. There is also a probability if not a possibility that the contractor would have underpriced the external foul drainage as he is sure that these shall be omitted and may have overpriced other sure items of work. This will create a double loss to the Employer and it would be difficult to prove the second loss.

1.1.11 Alternate Dispute Resolution (ADR)

ADR (median, Arbitration, Conciliation, Adjudication, Med-AR, and Mini-Trial etc.) is an indirect admission of construction irregularities, failures etc. This mechanism is meant to sort out the problems that arose owing to such disputes after the event has taken place. Technical Auditing with proposals for improvements is trying to stop or minimize such problems.

1.2 OUT-SMARTING BY CONTRACTOR

These are general guide lines based purely on experience and great care has to be taken for the application of the following as these can be considered a precedent. It is always better to discuss with the higher authorities and get their approvals. Some of these are in conflict with certain court decisions.

It is noticed that in certain contracts the Contractors are aware that certain items shall be omitted or changed during the execution. These can be as small as a protection slab over the drainage pipe or much bigger proportions that even the BOQ (with approx.. quantities) status as a uniform pricing document is under question. There were instances when the Final quantities were inserted instead of Tender quantities, the original lowest tenderer is no more the lowest giving serious doubt of the whole tendering procedure in order to obtain "value for money". Certain courts held that BOQ Rates should take precedence in addition, omission or pricing of additional works, may be the Courts were convinced that these Rates are not numbers have undergone analysis by Consultants before the award and hence considered to be reasonable.

It is the duty of the Technical Auditor to first establish that outsmarting or unethical practices were in fact taken place in the pricing, before proceeding to these drastic actions. This is to be implemented on a case by case basis and not for global applications.

1.2.1. BOQ Rates are not reliable for the use as a base for changes

1.2.1.1 Non use of available BOQ Rate as a base for pricing the varied work when the original BOQ Rate is very high.

1.2.1.2 In a project it was noticed that the capacities of the Air Conditioning equipment were changed during the implementation. A general pattern of under pricing was noted in the equipment that was in tender and subsequently omitted. This resulted in a substantial increase in the Variation. A new pricing mechanism was established on the basis that actual change in prices between the present capacity to that of the original capacity with the necessary Overhead & profit were agreed as Variation.

1.2.1.3 The Electrical cables were drastically changed during the execution of a project. While pricing the variation, it was noted that the cables were priced very high (around 300% of the market Rate), which resulted in a high amount for this Variation. This led to many discussions and finally it was agreed on the similar basis as above. All cables at tender and actual were priced with market Rates. The difference was given as Variation.

1.2.1.4 In another project the Contractor has under quoted the entire Electrical items. During the execution the entire Electrical items were changed owing to Service Authority's requirements. The Contractor priced all Electrical items as Variation based on market prices, which were accepted by the Engineer as these were new items, not expected at the time of tender. The End result was, the prices of changed much lower size cables were very much higher than the higher size cables in tender. The new prices of lower capacity distribution boards were very much higher than the higher capacity ones in tender. As an amicable solution, BOQ prices were given to Distribution Boards as the capacities were not increased so there is no justification to higher prices as these were not new items. Similarly the price of the new cable cannot exceed the price of next higher grade available in Contract. The price of 25 mm cable cannot exceed the BOQ price for 35 mm cable as the specification remains the same.

1.2.1.5 Non-feasible BOQ Rate (in relation to the given PC Rate) in Tender, or manipulation of PC Rates. The Engineer has to identify the PC Rate items and make necessary Arrangements with the Employer in order no manipulations can take place. A Mathematical Formula with only One Variable (Purchase Price) which is given in the Annexure 3.5.4 can be used for adjustments.

Very small quantity in tender, which were priced very high and the quantities increased drastically during implementation. Such cases to be noted during the analysis of tender and the Contractor should be made aware that such Rates shall be applied to tender quantities only. Additional quantities to be based on new reasonable Rates.

Addition of work after award

During the post-tender negotiation, owing to budget restrictions certain items were removed from the lowest tenderer and the project was awarded. These removed items were not analyzed during the award as these were not in the scope. It so happened that these rates were very high when compared to others. The removed items were again given to the Contractor as Variation, resulting in a heavy loss to the Client. This is a bad practice. Scope to be clear. Variations should be avoided. If there is a possibility for that item of work to be done later, it should have under gone some analysis and the Variation to be based on such adjusted/reasonable Rates.

1.2.3   Relying on the Contractor staff, conflict of interests

There is a tendency for the Consultant to rely on the Contractor land Surveyor for all levels as some Consultants do not have Surveyors. There are possibilities for cheating and the amounts are huge in Roads and Parks or projects with heavy earthworks.

1.2.4 Non-Contractor co-operation

There are instances where Contractors tend to request for extension of time for a slightest of a problem in order to cover their own delays. Recent case studies such as the ruling that Float belongs to the project, whoever gets there first can have it, Indicated that the Contractor also has to play an active role in the successful completion of the project.

1.2.5 Provisional Sums:

Provisional Sums is considered to be the prescription of the Tender documentation, meaning zero value is the best tender document. In practice, these have come into existence and in some Contracts there is a big percentage (sometimes in excess of 25%) of the total cost. There is a place for the Contractor to insert his mark ups for such provisional sums. Some Contractors do not price this and forget conveniently, sometimes arguing that they are unaware of the scope of such work. Provisional sums are generally not considered in details in the Tender Analysis for the award as there is no guarantee that the same shall be used. In the end when these amounts are used by external parties, the main Contractor claims a markup, sometimes no input at all or very little, if any. This is a serious case in Contracts where substantial amounts are given under provisional sums. One way of mitigating this kind of play is to analyze the Provisional sums in details, whether the same shall be used in full or part, or not a all, the contractor's markup etc. before the award. Experience has shown that the attitude of the Contractor before the award (more willing for compromise as there is a possibility to lose the project altogether) is much more different after the award-often termed in the construction circles as a different cattle of fish.

1.2.6 Play with Specification

In the present day market, equipment and materials are available meeting the same specification but with wide price variation depending on the country of manufacture. Even products from one country assembled in another country has a big price difference. (made & assembled in Japan Vs made in Japan & assembled in Malaysia).Needless to say that there is room for play in such situations.

1.2.7 Tender Design is not applied to construction

In certain Design & Build projects, particularly in steel structure projects, there is a tendency for the Contractor to submit the price with certain parameters, which is some form of over design. During implementation, the correct design is used. As the Consultant is more interested in the structure under construction, meeting the technical requirements, there is a tendency to ignore the original design which was used for pricing purposes.

The possibility for similar situation is in other constructions, say for example Swimming pools. There is a wide difference in the criteria for swimming pools as these can be for a villa, a school or even for Olympic

1.2.8   Use of Mathematical Formula for the weight of steel instead of actual weight

Payment for Re-bar weights in re-measure Contracts is generally based on a Mathematical formula as it is practically impossible to use the actual weight. There is substantial saving in projects where heavy reinforcements are used, if there is possibility to calculate the actual weight as there are underweight (relatively higher Carbon in the composition) steels are available in the market and the same is used.

1.2.9   Use of excavated soil instead of Sweet soil in Parks

There is a requirement for certain depth of Sweet soil under the vegetation. Some parks do have a series of mounds. There is a possibility to fill these mounds with excavated materials and then the required depth of Sweet soil, but billed the entire fill as Sweet Soil. This is a double loss to the Client as the excavated earth is claimed as disposed off site, whereas the same is used and claimed for filling.

1.2.10 Road Pavement thickness not meeting the requirement

Some Road or other pavement construction while meeting other requirements do not meet the specified thickness or the reduction is not within the accepted tolerable limits. Some Engineers totally reject same and ask the Contractor to remove and re-lay as per requirements. Some Engineers accept same with a reduction in price for that particular item if the same can be used as the same meet other technical requirements. Technical Auditors to check these kind of work as there are possibilities for malpractices.

The same is applicable for Interlock tile as thickness can be 60 mm but claimed for 80 cm as both on the surface looks alike.

1.2.11 Manipulation of Invoices.

In certain Contracts, particularly repetitive nature, say housing schemes, there are possibilities for changes, say for example, changes to ironmongery. Some Contractors with the help of suppliers inflate the price of items, where there was increase in numbers and reduce other item to make the final figure the same. The total price in the Quotation and the Invoice is the same, but detail pricing is manipulated. The Employers can lose a lot in this type of fraud.

1.2.12 Double Handling/Haulage

Double handling is generally not allowed in contracts. However, in certain projects particularly in build up or congested areas, there may be a necessity to pay for double handling. There are possibilities for contractor frauds, therefore, a correct system to be established in the project and the same to be followed,

1.2.13 Manipulation of PC Rates

The PC Rate for the supply of materials/equipment are given in the Contract in order to bring some sense as materials and equipment are available in the market with big price variations, although meeting the same specifications. This may be due to the country of origin. Sometimes products from the same country but assembled in another country also has a price difference. Sometimes Contractors front load certain items of work and under quote items of work which are done during the later part of the Contract. There is a possibility that these kinds of front loading are not considered in subsequent adjustment of PC Rates. An example is given below to illustrate the point.

In a major housing project, a BOQ item is given as follows:

Supply and fix Ceramic tiles to the floors as per specifications. (PC Rate for the supply of tiles is 15/= per m2. The same was priced as 17/= per m2. The accepted over heads and profits for the contract is 12%.The actual purchase price for the tile is 10/= m2. The Contractor has given a savings of (15-10) = 5/= per m2. This was accepted by the Engineer as it looks alright. The detailed study of the case resulted as follows:

This BOQ Rate is not feasible as the BOQ item includes for wastage,labour, plant, other materials and over heads & profit (12%). This will result in a negative value for labour & plant and the same should have been priced elsewhere. So a new method has to be employed; using a mathematical formula

$$\frac{1.03 \text{ X M} + k}{Y} \quad \text{x} \quad 100 \text{ as there is wastage.}$$

where the wastage is 3%, M is the Purchase price, K is the labour & plant, and other materials Y = 100 =-Overheads & profit), the new Rate arrived is 11.15/m2.

Therefore, the savings to be

(17—11.15) = 5.85/m2 and not 5/= per m2 as above.

1.2.14 Alternative Offers

In some tenders, there is a provision for the submission of Alternative Tenders in addition to the specified one. There is a tendency for the Employer to accept the Alternative tender without studying in detail owing to lesser final value. There are possibilities for the Tenderer to submit such offers with lot of qualifications resulting in serious difficulties during implementation.

1.3 OUT-SMARTING BY CONSULTANT/ EMPLOYER

1.3.1 Delay in payment

Delay in payment is considered to be the biggest problem for the Contractor as it creates a chain reaction as the contractor is often unable to meet the payment requirements for both staff and the material suppliers.The Conditions of Contract usually specifies the payment to Contractor as within certain no of days upon certification by the Engineer. The Conditions of contract also stipulates the action that the Contractor can take such as slow down or stop work in the event of non-payment. This action cannot commence on day one of the expiry of that period specified but after a further grace period, and another notice period for such action to take place. That means the Contractor cannot slow down or stop but has to wait further period. Some Employers do not pay on the due date but do so during the grace period, making the initial payment date meaningless. This is purely Contractual as it is the Contractor's risk, but I do not know how the courts will react if this kinds of situation prevailed as a rule rather than an Exception in a Contract.However, the Author feels this is immoral on the part of the Employer.

1.3.2 Selective removal of high priced items as Variations.

Some Employers remove high priced items of work from the scope as the Engineer has the Contractual provision to do so. This will reduce the Contractor's anticipated profit. These changes usually falls within the allowable limit for additions and omissions and the Contractor is unable to claim any loss of profit.

1.3.3 Large volumes of additional works given as Variation for items of work which are priced low in tender.

Some Employers give lot of additional works on the items which are priced low for the removal of high priced items of work. This affects the profit of the Contractor.

1.3.4 Unrealistic additional work given during the Contract period without any extension of Time

Lot of additional works are given without any extension of Time. It is sometimes impossible to complete such additional works during the Contract period making the Contractor to accelerate some portions of the works. The Employer refuses to give extension of time with associated costs or acceleration costs. In certain contracts, there is a provision for the Client to recover additional overheads from the contractor if the final contract price is in excess of certain percentage of the original tender price. Sometimes, Engineers apply this across the board without any consideration for additional staff that were employed owing to such additional work.

1.3.5 Unrealistic use of certain Clauses of the Conditions of contract or other provisions.

There is always a provision in the construction contracts, that the work and materials to be to the Engineer's approvals. Some Engineers misuse such rights which lead to losses to the Contractor.

1.3.6 Insitenance on the use of certain company products and sub-contractors

Some Clients insist on the use of some selected company products or sub-Contractors in which the Clients have a vested interest. There is nothing wrong if such cases were already in the tender documents. The problem is, it is imposed as some sort of thuggery after the award and there is a possibility for the contractor to Lose in such cases,

1.3.7 Taking more time than normally required for Approvals, Payments etc.

Engineers take unusually long time for approvals or for giving the required information to the Contractor. The Conditions of Contract always in favour of the the Engineer. For example, the Engineer can take 28 days for approvals, some of which can be easily done in a couple of days.These kind of delay in approvals delay the overall project and the Contractor is not in a position to claim.

1.3.8 Intepreting the Contract Conditions to the advantage of the Employer. Such as Interpretation of Circumstances that can be foreseen by an experienced Contractor.

As there is no clear definition of an experienced contractor, sometimes the Engineers feel that the Contractor should know better as he is supposed to build for the purpose and the Engineer's liability is limited to skill and care pertaining to their profession, There is a tendency for the Engineer to pass his mistakes to the Contractor.

1.3.9   Advance payment : both time of payment and subsequent deductions for same.

Advance Payments are usually given to Contractors by the Employers in order to get a competetitive price for the work.The Contractor also gives such a good price in anticipation of money early.Ideally the Advance payment to be made within one week of issuing the of Letter of Acceptence. In practice Advance payments are given much later, this affects the Contractor financially as he has to find financial resouces upfront,which was not the intention at the time of tender.There is another issue,the recovery of Advance payment.Some Employers base the reduction on value of work done,while others go by the periods elapsed. The Contractor shall be in serious financial difficulty in the event the Advance payment is given late and the recovery of same is based on time elapsed.

1.3.10 Rejection of Contractor entitlements owing to the mistakes in the documents/ drawings for not finding the mistakes and bringing to the notice of the Consultant during the tender stage.

This is a major issue particularly in Re—measure Contracts. The General understanding in such type of Contracts is the Quantities shall be re-measured upon completion and any items which are not measured is to be done by others.For example in the tender drawing a boundary wall is shown,but not measured. The understanding is that the Employer does not want to do the boundary wall now,or may be done by others.Some Employers in Re-measure Contracts,for which the Contract documents are prepared by the Employer (or his agents on his behalf) insist the Contractor to do the work,in this case the boundary wall without any additional payment as the Contractor has not notified this discrepancy during the tender period.This is not fair as this is not a Lump sum Contract,where the Drawings take precedent and the tender documents were not prepared by the Contractor,so he is not responsible for mistakes or ambiguities.

1.3.11 Delays in issing the Taking over certificate & Defects liability Certificates

Delay in issuing the Taking over Certificate affects the Contractor as he is unable to get one part of Retention and usually there will be application of delay Penalties when Contractor payments are processed after the original completion.

Delay in issuing the Defects Liability Certificate affects the Contractor as he is unable to get the second part of the Retention. Usually Performance Bonds are tied to the Defects Liabilty Certificate as such certificate is the final completion of the project. The contractor is unable to get his performance bond. This is a serious issue in some parts of the world as banks are insisting to deposit the full amount for Performane bond.

1.3.12 Imposing rules,regulations and procedures which are not given in the Contract documents.

Some unethical practices are followed by the Engineer/Client. For example,the Engineer takes the full advantage of the Contractors position that the Contractor has to build for the purpose. Certain Design & Build project do not follow the FIDIC Silver book standard of passing the entire risks including underground conditions as this will involve inflated prices irrespective of the underground conditions. The Employer gives the Soil Investigation report as a part of the Tender documents. These are given with lot of qualifications. When the project is implemented, the soil conditions at site were so bad that improvements are required. This additional cost is rejected by the Employer, stating that the given Soil Investigations were given with many qualifications and it is the responsibility of the Contractor to check the correctness of the given soil data. This is in conflict with certain Contractual Clauses such as

Clause: Contractor's General Responsibilities:

The Contractor shall, with due care and diligence, design to the extent provided for by the Contract, execute and complete the works and remedy any defects therein in accordance with the provisions of Contract.

Clause: Inspection of Site:

The Contractor shall be deemed to have inspected and examined the site and its surroundings and information available in connection therewith and to have satisfied himself so far as is practicable, having regard to considerations of cost and time before submitting his tender, as to: The form and nature thereof, including the sub-surface conditions

Clause: Adverse Physical Obstructions or Conditions

If, however, during the execution of the works the Contractor encounters physical obstructions or physical conditions, other than climatic conditions on the site, which obstructions or conditions were, in his opinion, not foreseeable by an experienced Contractor, the Contractor shall forthwith give notice thereof to the Engineer, with a copy to the Employer.

The Author feels that in a situation like this the Courts shall favour the Contractor as although the soil Investigation was given with lot of qualifications,the Contractor is not in a position or not allowed to do the detailed investigation and left with no option other than using the Employer's given soil data for his design.

If the intention of the Client is to pass the complete risks to the Contractor,then Clauses as given above,which indicate the sharing of the underground risks if and when available should have been removed from the Contract.

The Author feels that the fair solution to these kinds of problems (as these are underground) is to pass the entire risks to the Contractor and get the inflated price irrespective of the presence of such risks or inclusion of reasonable provisional quantities at tender for substructure,which is subjected to re-measure upon completion.

1.3.13 Not awarding extension of time before completing the project

The Engineers do not understand that the Contractors do have commitments to the suppliers and other service providers.The Engineers usually do not give the Interim Extension of time which the Contractor is entitled as per the Conditions of Contract and wait until the final completion.This affects the Contractor financially as penalties are imposed on payments which are processed after the original completion date.

1.3.14 Undue delays in finalizing the Final Accounts of projects.

Finalization of Rates particularly Varied work or variations take a very long time.The Contractor is generally paid on account payments which are at best only a fraction.This affects the Contractors Cash flow as these works were already done and the contractor has to make payments to the suppliers.

1.3.15 Sometimes ,circumstances,which force the Contractor to accept what ever is given by the Employer

Sometimes the Contractors are in such a bad financial situation that they are prepared to accept whatever given by the Employer. This may be the result of delay in the Employer's approvals of Contractor's entitlements are so late that there are lot of financial pressure inserted by his suppliers.

1.3.16 Inappropriate design,which leads to abortive work for which payment is not made.

Sometimes,there is a possibility for the correction of Engineer's design faults.These often results in work done by Contractors as per the original design,subsequent removal and re-done as per correct requirement.It is very difficult to get approvals for Abortive works resulting from the Engineer mistakes. The Engineer goes back to the theory or the cover that the Contractor has to build for the intended purpose, etc.

1.3.17 Forcing the Contractor to do "As Built Drawings"

The preparation of "As Built Drawings are under the scope of the Engineer as there should be some kind of Check & Balance. In practice all the "As Built Drawings are prepared by the Contractors and the Consultants get paid for this work.

1.3.18 Employer acts as Engineer

In certain Government Departments or other big co-operates, the Director or Head of the Department also acts as the Engineer. This is not correct as in most ,if not all Conditions of Contracts,the Engineer should be impartial and to be fair. It is impossible to be impartial or fair when there is a vested interest for the concerned party.

1.3.19 Unethical practice,requesting Overtime for some work.

Some Consultant's staff force the Contractor to pay overtime by arranging certain work after the normal working hours. As this Overtime payments are deducted from the Contractor payments,the Client also not that much worried.This is an unethical practice by the Employer staff.

1.3.20 Use of Unrealistic amounts for Contractor Delays

It is an accepted practice to deduct some amounts per day for the delay solely on the part of the Contractor as the Client is getting the product late, thereby losing the return from the investment. Some Clients charge unrealist amounts.This Liquidated damages to be reasonable.European Courts have always of the view that while Liquidated damages can be admisable but are very reluctant to enforce a penalty,which sometimes has no logic and some form of deterent.

Some Clients do charge unrealistic amounts and in addition sometimes charge the Consultant's supervision for the extended period,arguing that such supervision was required owing to Contractors' fault. This has to be studied on a case by case basis.

1.4   OUT-SMARTING BY CONSULTANT/ CONTRACTOR

1.4.1   Issuence of completion certificate before actual completion.

The general procedure is that the Contractor gives notice upon completion of the works and after due consultations and inspections, the Engineer issues the completion certificate if the project is substantially completed.It is also an accepted practice for the Contractor to attend to small corrections which are known as snags during the Defects Liability period.The understanding is that , upon substantial completion,the product can be used for its intent and purpose inspite of snags.The Employer is contractually entitled to charge pre-agreed amount for delay on the part of the Contractor.Some Engneers issue the Taking over Certificate before the works are actually completed in order to avoid delay payment by the contractor and allow to do large amount of remaining portion of the works as snags..The Dictionary meaning of the word snags is unexpected obstacles or drawbacks and does not imply the remining works or portion of the works.In such situations,the Engineers supervision is usually paid by the Contractor.The end result is follows:

a)   The Engineer gets his supervision fees

b)   The Engineer's design faults, or corrections which may necessitate such extensions shall not be highlighted.

c)   The Contractor escapes without paying delay penalty

d)   The Employer loses his right to charge the delay penalty

e) Much reduced Defects Liability period for the works done after the issued Taking over date as the Defects Liability period is based on the Taking over certificate date.

1.4.2 Creating a "Time at Large" situation to escape Contractor delay deductions

It is an international practice to apply Liquidated damages or Penalty in the event of contractor delays as the Client is not getting the return on his investment. In some contracts,the Contractor and Engineer makes or bring the project to a "Time at large "situation by giving works after the original completion date when there is a contractor delay.These additional works are part of the contract as Provisional sums. In theory, the Engineer can omit partly of in full the Provisional sums.The question is, why the Engineer took such a long time(sometimes very much later than the original completion date) to decide on these provisional sum items. In the end,the Engineer is compelled to give Extension of time and to make the Client happy, he will recommend extension without Associated costs. The end result is

*   The Engineer gets his supervision fees as the work has to be supervised.

*   The Contractor escapes without any delay deduction

*   The Client losses by paying additional supervision and foregoing delay penalties.

1.5 CHECK LIST

A brief check list for common small and medium size projects is given below:

a) Amount and approval of the budget:

b) The scope of work .Is it clears both in terms of requirements of Employer and End User, if any. Is the specification of materials and equipment clear enough, in order to give the level playing field for all intended Contractors.

c) Have all Government Service Authorities requirements are met in the proposed scope. Both in terms of General approvals to Technical approvals.

d) What are the Risks in this project? What actions that are taken to mitigate such risks.

e) What is the most suitable procurement method for the project? What are the minimum documentation that are required for tender as per Technical Auditing Standards?

f) Has the Contractor appointed with the due process-Tender analysis,recommendations by the Engineer etc. in order to obtain the best possible Value for money.

g) Has sufficient Post—Tender negotiations taken place in order to make the scope clear to the appointed Contractor, the availability of resources with the Contractor, possible discounts from the Contractor.Completeness of the Contract documentation.

h) Base line programme—Can this is acceptable in general, in relation to the project Scope, time and Contractor resources. Procedures for monitoring and correction during implementation. This also one aspect that has to be given consideration during the post-tender negotiations and before the award.

i) The general procedures to be followed during the execution of the project, both in Terms of construction and finalization of project accounts, such as but not limited to

- Procedure and Agenda for the proposed site meetings

- Procedure for Approvals

- Period for Engineer approvals

- Site logs

- Safety and quality procedures

- Final measurement procedures, target dates for re-measure projects.

- Finalization of varied or additional works.

- Finalization of Contractor entitlements

- Project close out procedures.

j) The payment arrangements for the Contractor, starting from the Advance payments, if any, and their method of recovery, subsequent monthly or interim payments, and Final payment. This one area where the Engineer has to use his knowledge and experience to have a suitable arrangement, which is beneficial to both parties and maintain a very high level of confidence. There is a possibility for this action to take place during the post-tender negotiations.

k) Provisional sums. This is the prescription of the tender document, meaning zero provisional is the best. So it goes without saying that the amount under this to be the minimum

l) Insurances—Does the Contractor have sufficient Insurance cover, both for the intended project, equipment & labour.

m) Does the Engineer have Professional indemnity Insurance, is the same cover for Professional negligence?

n) Sub—Contracting arrangements –Are these satisfactory ?

o) Is there provision for Technical Auditing in the Contract?

p) List of procedures that are to be followed for Mega projects

- Feasibility studies

- Viability Studies

- Status of required materials, possibilities for the escalation of prices of such materials

- Is there a necessity to have public debates regarding the project? If so had it taken place.

- Auditor's report

Note: **The above are basic requirements for small and medium size projects and each project to have its own list of requirements. The above list does not cover Mega projects involving Government and private parties as these are expected to have feasibility, viability studies and there is a requirement for Technical Auditor to play a role before the award of such projects.**

2

CONSTRUCTION FRAUDS

Frauds can be considered to be Decit, Trickery, Sharp practice or breach of confidence, perpetrated for profit or gain, some unfair or dishonest advantage. The broad legal definition can be –There must be a deliberate misrepresentation of the product condition and actual monetary damages must occur. All these are applicable in some form or another in construction frauds.

Construction is not immune to Frauds and like any other industry; construction too is affected by frauds. Frauds shall occur at various levels such as workers, main Contractor, sub-contractors, Client staff etc. The ultimate effect which is an unnecessary loss is felt by the Employer.

2.1 Types of Contract Frauds

2.1.1 Bid Rigging

The pre-qualified contractors raise prices of the intended project. The Employer who has solicited competetitive bids in order to procure the project at the lowest possible price. Essentially,competitors agree in advance who will submit the winning bid on a contract, that is being let through a competitive bidding process. So on paper, everything looks alright meeting both the requirements of transparency and fairness.

Bid rigging includes colluding for bid suppression (agreement among bidders to withdraw or suppress bidding) complementary bidding (courtesy bidding purposely high) bid rotation (bidders take turns being the lowest) and subcontracting (a low bidder will agree to withdraw his bid in favor of the next low bidder in exchange for a lucrative subcontract that divides between them the illegally obtained higher price)

Frauds are of various types.-Removing materials, tools, equipment etc are considered to be non—technical frauds, while manipulation with specification, drawings, payments and other unethical work can be termed as Technical Frauds.

2.1.2   Price Fixing

In this case, competitors agree to raise, fix, or otherwise maintain the price at which their goods or services are sold. It is not a necessity that competitors agree to charge exactly the same price or that the every competitor in a given industry to join the conspiracy. Price fixing includes actions that establish or adhere to price discounts, hold price firm, eliminate or reduce discounts, adopt a standard formula, which does not favour the Client, for computing prices, maintain certain price differentials between different types, sizes or quantities of products. Adhere to a minimum price or fee or price schedule. Fix payment terms.

2.1.3   Product Substitution

This involves introduction of equipment or materials from other sources that also satisfies the specification. There is a wide price variation for the equipment and materials produced by different countries but meeting the same Engineering specification. The Contracts specifically calls from one particular country, where as the contractor provide same from a different country, thus making a unjust profit.

2.1.4   Bribery

Bribery is said to take place when a person or company directly or indirectly gives, offers or promises anything of value to any public official or person who has the authority in return for some monetary gain. Bribery involves influencing any official or persons with authority to commit or aid in committing, colluding in, or allowing any fraud, creating an opportunity for the commission of any fraud, or inducing a public official to do or omit to do any act in violation of the lawful duty of such official per person.

2.1.5   Kickback

A kickback is money paid for referral of business for a contract,without the knowledge of a customer and without the customer's best interests in mind.

2.1.6   Conflics of Interest

A conflict of interest to takes place when an employee in a decision making position has a direct or indirect interest, particularly a substantial financial interest, that influences the individual's ability to perform his job and responsibilities. A situation where an official's private interests may benefit from his or her public actions.

2.1.7   Very high content of Provisional sums.

Generally speaking, any work which is not included in the original tender and done after the award is definitely going to cost more than if the same was included in the original tender. Some or most of this work shall be done by others and the main contractor will get his mark up with little or no input.Favourtism, inside dealings and phantom companies are possible if the sub contractors who do these works were not decided at the time of tender. I can still remember when I interviewed a QS and asked him do you see any abnormality in the Tender price when the Provisional sums were around 30% of the total. The QS replied no, again when I mentioned 30% Provisional sums, he said ,I have seen 50% and 30% is alright and there is nothing abnormal. The Author feels high provisional sum is an irregularity bordering Fraud as it may lead to Frauds.

2.1.8   Hiding design corrections, lays on the part of the Engineer

There is a tendency by the Engineers to hide their faults such as correction of design during construction, delay on the part of reviews, approvals etc. in a timely manner. In such instances, it would be advantages for the Engineer and the Contractor to convince the Client to grant extension of time without cost. Such requests are generally accepted by the Client. The end result is the Client gets the project late and loses the anticipated return on the investment. The provision of liquidated damages allowed in the Conditions of Contract for such an eventuality is not enforced as Extension is already granted. This is an avoidable loss to the Client.

2.1.9 Change order manipulations

Omissions are priced low while additions are priced high. Work items are shifted to Day work situations. There are possibilities to remove from in the original contract under quoted items, or items where there is a sudden increase in prices after the award and treat same as varied works for the benefit of the contractor.

2.1.10 Errors in measurements in a Re—measure Contracts

Over measurements of completed works, which is a loss to Client.

It is often difficult, if not impossible to find out Technical Frauds by Financial Auditing.

It is rather unfortunate that Government funded Construction projects are still and continue to undergo Financial Auditing only in many countries, particularly in developing countries.

It is a fundamental requirement to have Technical Auditing in construction to elevate the confidence level of the industry.

2.2   Non-Technical Frauds

Removing of materials either from site or divert the materials belonging to one project to another,

Removal of equipment

Manipulation of labour—Increase the no. employed/hours worked in the Time sheets

Manipulation of Plant—Increase the no or the hours of equipment in the Time Sheets

Manipulation of Prices of materials—Invoices from Fhatom companies

Manipulation of Quantities—Invoice more than what is supplied

Under production, or idle labour & plant.

Over payments to Contractor—Work measured more than what is done.

It is difficult to check these kinds of faults as the Employer cannot employ people to check these kinds of activities at the construction sites. These kinds of activities are mostly in Cost plus Contracts with an open book. The easiest way is to stop this kind of Cost plus contracts, where there is no incentive for the Contractor. On the Contrary, the Contractor gets paid for everything, more spent it is good for the Contractor. All Contracts to have some form of competitiveness ,to arrive at the Contract Price along with clear scope, proper Tender and Contract documentation.

For further reference and case studies regarding Frauds in construction read "How to win and Manage Construction Projects by the same author.

3

TECHNICAL AUDITING OF CONSTRUCTION

3.1   INTRODUCTION

Construction is a necessary industry with lot of potential for growth owing to the needs of the community to have shelter and other infrastructure to live in the modern world.

Certain studies indicate that in a couple of decades as a first time in the planet more than 50% of the population shall live in cities. This creates a situation, where village people shifting to towns in large numbers. This has already commenced in China to a very great extent. Therefore construction in one form or the other contribute substantial portion of the expenditure and thus creates employment opportunities.

The Construction Industry has a poor rating as an industry owing to many irregularities on the part of Professionals and Clients coupled with Frauds, litigations, and no universally accepted standard of Technical Auditing procedure. The court cases in the industry are only comparable to that of extra-marital cases. No wonder that the International investors are not considering construction as an option even in the present very bad investment climate. Something has to be done, and that too as early as possible.

Background

The majority of the Construction projects in the developing countries are done by a process of Budget allocation, Design by an appointed Consultant or in house and the construction by a contractor. The Consultant and the Contractor are selected through a formal tender process. The appointed Consultant could then be responsible for Only the design phase, in which case a different consultant would be appointed to Review the design and supervise construction, or Both the design and construction supervision.

During construction, a team from the concerned Department attends monthly site progress meetings in order to keep track of progress of the works and encourage adherence to the project specification. This kind of arrangement had in the past both success and failure.

Upon substantial completion of the project, the Government Audit Department may conduct **a Financial Audit, but not a Technical or Performance Audit to see whether the concerned Department has obtained what it paid for, is carried out.**

On completion of Construction, a team from the Department along with Consultant (if any) inspects the project identifying deficiencies and defects, which the contractor has to correct before demobilizing and within the Defects Liability Period.

This post-construction inspection is essentially a visual evaluation and does not usually identify whether the materials used in construction were what were specified or the quality of construction complies with design specification. Technical or contractual weaknesses and the construction defects that are not identified at that time frequently result in premature distress which may occur after the stipulated Defects Liability Period, making it difficult for the concerned Department to have adequate recourse to claim from the contractor. The situation is made worst, particularly in the developing world as Contractors do not provide Guarantee (or an insurance cover in the event of failure) of the product for a further period of say 10 or 15 years. Insurances are generally provided only upto the end of Defects Liability Period.

The use of Formal Technical audit during construction and upon completion allows the concerned Departments to identify whether the parties involved in the construction Contract were correctly constituted in the first place and the Client is getting the best possible value. This involves a detailed assessment by suitably trained professionals in terms of compliance of the materials and construction with the design specification, than the post-construction inspection,

The primary objective to have the Technical Auditing Protocol and other guide lines is to give a general standard arrangement for such audit practice.

Like in any other Industry, construction is not immune of frauds in spite of the presence of various Cost Engineering professionals and Institutions, whose ultimate objective is to give the best possible value for money.

Owing to the complex nature of construction, the industry as a whole is infested with all kinds of problems and comparable only to that of extra martial cases. This may be the reason why international investors are leaving construction as an option, particularly in the present economic environment where the established forms of investments are subjected to many kinds of hereto not known or seen risks.

Some forms of Construction Auditing, which is mostly financial is still done, particularly in Government establishments owing to public Accountability. Construction Frauds were and continue to be highlighted practically in all the countries by their respective Auditor Generals. The estimated percentages of construction Frauds vary from country to country but there is no denial of its existence whether developed country or otherwise. This is still come under the scope of Chartered Accountants although it is plain to see that it is an outdated practice owing to the complex nature of construction. It is rather unfortunate that Cost Engineering professional Institutions have not made any tangible steps to correct this situation with say for example the introduction of Accredited Technical Auditors similar to that of Accredited Mediators, Expert witnesses, Arbitrators etc.

The quality and quantity of the findings is directly proportional to the level of sophistication of the Employer and the contractor. While the present focus of the audit may be on over charges as it is primarily done by Accountants, the Technical Auditor being an Engineer with great experience in constructions is in a suitable position to make suggestions related to best possible project procurement, project controls and industry best practices.

For example, the Government Department, which started a new project, is unfamiliar with the Construction best procurement and management methods and complicated construction payment applications and may lack suitably trained staff to adequately support the project and review invoices, they may be completely reliant on the Contractor for project management and controls. The Contractors usually focused on maximizing profit and using the terms of contract to their benefit. This creates a situation where the owner has a high risk of overpayment to the Contractor and costs and time over runs. The Technical Auditor can help to protect the

Employer by including their findings, a draft of stronger and clearer language for future contracts, which anticipates and addresses potential problems. The Auditor can also help the Employer by making recommendations to strengthen controls, and by identifying unallowable or duplicated charges, overcharges and other errors.

Auditor Generals are generally given wide powers, in certain countries (such as India) even to the extent to advice Legislature. It is about time that Construction Auditing to be conducted by suitability trained and accredited Engineering Technical Auditors.

My personal opinion is that the Employers should spend at **least One Dollar for every Thousand Dollars** spent, to see whether the money is spent properly in relation to "Value for Money"-the best possible value for money spent. It is a moral requirement in the case of Public money.

It is time that Engineering Professionals in general and those involved in the costing side of construction realize that they have a moral and professional duty to the society and country at large to introduce a set of well defined Proposals for Construction Auditing to give Value for the money spent on such constructions.

There is a general reluctance for Technical auditing on the part of contractors and consultants.In fact they should welcome such an arrangement as the same shall give recognition to their companies as clean and prepared to give value for money.

There is a genuine concern by the contractors that Technical Auditing arrangements shall delay the Finalization of the project accounts.There is a necessity for all parties to co-operate and finalise accounts before the issuance of the Defects Liability Certificate.

3.2   WHY TECHNICAL AUDITING
IS REQUIRED

There is a necessity for Accountability, check and balance for the expenditure by the Government, which creates a requirement of Performance Auditing which is Technical Auditing of construction projects undertaken by the Government.

It has been noticed that substantial amounts are being spent by Government Institutions on repair, rehabilitation etc. on a regular basis after the Defects Liability Period (which is usually one year after the Taking Over) for many construction projects. There is no Guarantee of the product (or an Insurance cover for damages) given by the Contractor after Defects Liability period, which reinforces the necessity of Performance or Technical Auditing in construction projects.

Further, Construction projects carry lot of risks and there is a necessity to audit and make improvements as construction is a continuous and growing industry which involve substantial amount of public money.

The Technical auditor should analyze the cases and find out whether the fault of the Contractor is owing to greed or by circumstances which are beyond his control. We had cases, where some construction materials such as steel, cement, copper prices increased drastically, some even upto 300% resulting in the closure of some contracting companies. Solutions for such kind of problems to be suggested by the auditor.

The given scope of Technical Auditing, Activities of Technical Auditor, Protocol etc can be considered as basic guide lines for Technical Auditing for Developing countries.

3.3 SCOPE OF TECHNICAL AUDITING

The scope of Technical Audit, which is Performance Auditing is to evaluate the Technical compliance of the Project's documentation, work processes and the work in place. The objective is the concept of "Value for money, that is the best possible value for the money spent. The scope of which among others include the following:

1. Review the Original design, feasibility study, viability study, value Engineering and comparison to that of detailed design drawings to assess the net design variations, changes and justification of ensuring variation orders.

2. Review the type of procurement of the project, its compliance and suitability as the standard/best commercial practice.

3. Review the Contractors' means and methods, manpower, plant at mobilization and maintenance during the implementation period

4. Check the construction at the site,whether undertaken according to the specification and other test requirements.

5. Review the Project compliance with Operation and maintence, test requirements, warrantees,gurantees for the complete operation of the project in line with the objective of Build for the intended purpose

6. Review all Contractor payments, both Interim and Final, particularly regarding the unit Rate application and whether the best practices were followed in the pricing of Variations/ additional works, use of provisional sums, etc.

7. Review Contractors' Claims

8. Review Consultants' claims.

9. Make the necessary Audit reports indicating the short comings, savings etc.

10. Lessons learned and Recommendations for improvements in the future.

3.4 TECHNICAL AUDITING PROTOCOL

The Audit shall be carried out generally in accordance with the standards set out by the Auditor General of the state/country. It is advisable that these kinds of Auditing shall be performed by third parties from the Auditor General approved list of Technical Auditing Consultants.

The Auditing Team shall be headed by a professional Engineer with at least 20 years of experience in design, supervision and management of construction projects. The individual shall be referred to as the Auditor.

The Auditor may be assisted by Engineers specializing in the following fields who should have at least 3 years experience in their respective fields with emphasis on the country and are required to be Accredited as Technical Auditors

- Civil Engineering

- Mechanical Engineering

- Electrical Engineering

- Quantity Surveying/Cost Engineering

The size and composition of the team could be varied to suit the needs of the project provided,however,that the Auditor always heads the team and a QS/Cost Engineer is part of the team.

The Auditing team, types and Fees shall be subject to the Approval of the Auditor General. However, the Fees shall be made up of two portions namely a lump sum in the region of 0.1% of the budget and sharing of the resulting savings in an approved way in the region of 1/3 to the Auditor and 2/3 to the Client. In the event, the resulting savings is in excess of 5% of the Contract price, then the Auditor's Fee is only based on the Savings and the other initial lump sum is not paid.

Types of Audit to be performed

Construction Projects upto US $ 250,000/=
Final Audit

Construction Projects from US $ 250,000 to 2 million
One to two Intimidate Audits
Final Audit

Construction Projects from US $ 2-10 million
Two or Three Intermediate Audits
Final Audit

Construction Projects from US $ 10 to 20 million
Three Intermediate Audits
Final Audit

Construction Projects in Excess of US $ 20 million
Initial Audit
Four Intermediate Audits
Final Audit

Article One

All state funded construction projects should have the relevant provision in the Conditions of Contract to undergo the stipulated Technical Auditing procedure before finalizing the Project Accounts.The Contractors should welcome such an arrangement as the same will improve their ranking.

Article Two

The term "Construction project shall mean all capital projects dealing with Buildings, Roads, or similar constructions or technological improvements or other improvements using state funds.

Article Three

The Auditing Team, the type and number of Audits shall be decided by the Auditor General under the guidelines given in the annexure or whatever suitable as per the cost or the complexities of the project under consideration. All Fees for Construction Auditing shall be subject to the Auditor General's approval.

The Individual Auditing Team shall be appointed by the Director of the Department with due approval from Auditor General. The Technical Auditor shall directly report to the Client.

Under no circumstances will the Technical Auditor advice or issue instructions to the contractor or supervising consultant. Communications should be focused on seeking clarification or information regarding the project, and should avoid any interference with smooth implementation of the project.

The consultant, contractor and Client must make available to the Auditor any document as and when required by him and relevant clauses in the tender documentation/letters of appointment shall make provision for this.

In the event that there is a strong possibility that the original project completion cannot be achieved and in such cases the Engineer shall prepare a report called "Report on the scheduled completion date". The report shall detail the completed portion of work, its value along with the uncompleted or yet to be completed work, its approximate value, the reasons for such prolongations, the remedies for the earliest completion, recommendation, if any. All Conditions of Contract shall have provision for same.

Article Four

All Technical Auditing shall be conducted by approved external third party Technical Auditors. The Technical Auditor should familiarize and should clearly understand the scope and complexities of the project by reviewing all contract documentation including drawings, specifications and conditions of contract and site visits.

Article Five

The Audit start shall be stated in the Conditions of the Contract but not later than 30 days before the anticipated substantial completion of the project. The Audit report based on the Statement at Completion to be completed before the issuance of the Defects Liability Certificate.

Article Six

The Final Audit Report shall high light the short faults of the expected standard of the final project, the over payment and the savings generated by the Auditing with simple and clear terminology in order to have validity in the court of law in the event the contractor challenges same.

Article Seven

The initial audit shall be of two types, one for mega projects in excess of US $ 100 million. The initial Audit for such project shall be before the award and shall be specifically given based on the complexities of the project. The general guidelines for Initial audit for other projects are given below:

This should be carried out within 15 days and not more than 30 days after the mobilization.

The initial audit is carried out after the commencement of construction so that all of the correct procedures can be established from the beginning of the project. The first on site audit to focus on project management and construction methodologies such as

- Suitability of contract documentation, Route of procurement and award.

- Review of Consultant's control, approval procedures

- Suitability of site staff

- Knowledge of Contract, Site organization and communications

- Quality of completed work

- Suitability of Equipment & operatives

- Quality and suitability of the programme

- Site safety, Quality assurance procedures

- Suitability of the contractor's project management systems

- General attitude regarding project

- Payment arrangements

Article Eight

It is emphasized that all Mega projects are not the same. Each project should have its own guide lines as these involves substantial amounts of investment and the projects are expected to continue financially viable for a considerable period. Some General guide lines are given below, taking into consideration that the success rate of BOOT projects is around 20%.

Audit the truthfulness of Feasibility studies, adequacy of data, problems in predicting future, and experience on the part of the forecasters

Audit the Truthfulness of Financial and Viability studies in consideration of long period of study that are required for such projects.

Truthfulness of costs and benefits Forecasts

Psychological effects of over optimism

Political or Consultants influence to have the project implemented

The position of the project site or land, outright purchase, long lease or any inside trading for malpractices etc.

Is the above cost included in the studies?

Credit Risk or Equity risks to be studied.

The cost of the product or raw materials to be used for the project is assured in the long run and at reasonable cost?

The supply of energy at reasonable cost is assured?

The building materials are available at costs estimated?

The Contractor/operator has sufficient know how of the project. ?

The construction and other agreements are satisfactory?

Are the management personal are competent and capable for the project?

A stable political environment now and forceable future is possible in the country?

Does any equipment require overhaul or changes after the initial period of use?

Any chance of the equipment becoming obsolete?

What is the running and maintenance cost of the product, will it end up as a white elephant?

Article Nine

The intermediate Audit shall broadly audit the following

The work at site in relation to the programme

The Contractor resources

The measures employed for the measurement of completed work and the materials at site

Interim Payments

General correspondence systems

Interim Contractor Claims

Pricing mythologies for pricing variation/additional works

Procedures for the finalization of the project accounts.

Article Ten

The Final Audit shall broadly audit the following and submit an Audit Report
The truthfulness and compliance regarding the award of the project

The general progress of the work in relation to the approved programme and the provision of manpower and equipment by the contractor

Necessity and pricing of variations/additional works

Truthfulness of Measurement of actual work in relation to physical or approved "As Built Drawing"

Truthfulness of the use of Provisional and contingency sums

Operation and Maintenance manuals

Guarantees and warrantees

Test reports of work and materials

Truthfulness regarding the follow of Specifications

Regarding work, materials & equipment—Actual Vs Contract

Truthfulness of the use of escalation of prices of materials,if there is aprovision for same.

Contractor claims

Consultant claims

Quality Control: Inspection reports and confirmation all violations or variants have been rectified.

In Contracts where there is a target price and sharing of savings, these savings to be audited for truthfulness.

The Final product in relation to the anticipation at the beginning; Meeting expectation, more than expectation not Meeting the expectation.

Audit Report indicating the extent of the Final product in relation to the Contract, all short falls, unfair practices, if any, the resulting financial savings with details along with lessons learned and Recommendations

Article Eleven

All the Government Departments Contracts to have provision for Advance payment. This should be limited to 10% of the Contract price with the upper limit of US $30 million. The procedure for the award of such advance payment is as follows:

1. The Contractor should have commenced mobilization

2. Submitted the Programme of work as per Conditions of Contract

3. Submission of Performance Guarantee as per Conditions of Contract

4. Submission of Insurance as per Condition of Contract.

5. Advance payment Guarantee as per conditions of Contract and among others to include the following:

    The guarantee shall be free of interest and payable in cash on the first written demand in the manner ordered, without the contractor or any other person on his behalf or the bank having the right to suspend or delay payment or to object thereto for any reason whatsoever.

This bank guarantee shall remain effective until the advance payment has been repaid in full. The Advance payment shall be repaid through deductions from interim payments calculated and certified by the Engineer. Deductions shall commence with the first full month interim payment certificate after the contractor received the advance payment. The complete advanced payment must be repaid prior to such time when the accumulative payments under Interim Payment certificates reach about 90% of the initial Contract price. The Engineer shall determine the advance payment repayment period.

Article Twelve

Code of Ethics for Auditors

- Adherence to the Rule of Law

- Integrity and High moral values

- Objectivity and impartiality

- Diligence and Sense of Duty

- Keeping Professional Secrecy

Article Thirteen

Use of relevant Technology

The Auditors should take note of Technological developments that are taking place in the construction industry and should be in a position to use same if required.

3.5.1   AUDITING PROTOCOL—ANNEXURE ONE

FORMAT OF FINAL AUDITOR'S REPORT

1. **Project details**

Name of Project
Concerned Government Department/Entity
Engineer/Consultant
Contractor
Contract Price : US $
Contract Period: Days
Date of Commencement:
Date of Completion:
Actual completion:
Extension of Time Requested by the Contractor: Days
Extension of Time approved: Days
Final Contract Value submitted by the Contractor:
Details of additional Budgetary provisions, if any:

1.

2.

Total

2. **Auditors Findings**

1. The Professionalism & Truthfulness of the Engineer's Report on Scheduled Completion date in the case of extensions to original completion date.

2.

3.

4.

Audited Final Account: US $
Total Amount of Audit generated Savings: US $

Ref.	Description	Amount Claimed by Contractor	Amount agreed by Auditor	Savings	Remarks
1.					
2.					
3.					
4.					
Totals					

Amount of Savings agreed by the Contractor US $

Ref.	Description	Amount
1.		
2.		
3.		

4.

5.

- - - - - - - -

 Totals

- - - - - - - -

Total Amount of Savings not agreed by the Contractor and in dispute: US $

Main Reasons for the dispute:

a)
b)
c)

4. Lessons Learned

5. Auditors Recommendations

1.

2.

Signature of Auditor   Signature of Head Auditor

Name: Name :

Date: Date:

3.5.2 AUDITING PROTOCOL— ANNEXURE TWO

GENERALISED FLOW CHART FOR TECHNICAL AUDITING.

Planning

Decision to Conduct an Audit
Selection of an Audit type
Selection of Audit Team
Planning meeting
Confidentiality and Audit Results
Review of Project documents
Contact with Auditee
Audit Plan and other preparation
Audit questionnaire/Check list

Performance

Audit Protocol
Opening meeting
Audit Activities
Observation of works
Interviews
Document Review
Objective Evidence Compilation
Closing meeting

Evaluation

Identification of Findings
Determination of Corrective action

Documentation

Draft Finding report
Final report flagging any situation that might indicate
    fraudulent or corrupt practice.

Corrective Action

Implementation
Verification of Effectiveness
Closeout

Closeout

Closeout Memo
Audit File Closure

3.5.3 AUDITING PROTOCOL-ANNEXURE THREE—RECOMMENDED PRACTICE

All Government funded construction and maintenance projects are recommended to follow the following procedure

1. All construction projects to be graded for example:

    Grade One: Upto Us $ 100,000/=
    Grade Two: US $ 100,000-2 million
    Grade Three: US $ 2-5 million

2. All Contractors to be pre-qualified based on their resources and Departments to have list of such Contractors with their grades.

3. All project above the value of US $ one million to have the following as Tender documents as the minimum.

    One Volume for Conditions of Contract

    One Volume for Specifications

    One Volume for Bills of Quantities

    One Volume for Tender Drawings.

4. All projects in excess of US $ one million to have the following procedure for the award of project.

    4.1 The prequalified Invitees to be in excess of 12 nos.

    4.2 There should be atleast Five Offers

4.3 The Offers to be analysed with the following minimum criteria before the award

4.3.1. Reference

4.3.2. Tender Documentation

4.3.3. Tender Results

4.3.4. Tender qualifications

4.3.5. Detailed Tender Comparison

4.3.6. Feasibility of each individual item

4.3.7. Unit Rate Analysis

4.3.8. Day work Rate analysis

4.3.9. Detailed analysis of Alternative Offers in any.

4.3.10 Conclusion

4.3.11 Recommendations

5. If there is a Consultant

Method of appointment should be on the similar lines to that of Contractor

All Consultants to have Professional Indemnity Insurance and the same should cover

Professional Negligence. The cover to be above the average total value of project/s by Consultants per year.

The Technical Auditor is expect to see whether the correct applicable procedures were followed

3.5.4 AUDITING PROTOCOL-ANNEXURE FOUR—RECOMMENDED PRACTICE

MATHEMATICAL FORMULA FOR PC RATE ADJUSTMENTS.

Objective

PC Rates are being used in many projects and in some cases ,the given BOQ Rate results in a negative value for Labour& plant,indicating some front loading activities and hence there is a requirement to have a mathematical formula to make the necessary adjustments in the Interim payments and Final Accounts. The following are the guide lines to be used.

1. The BOQ Rate for the item is considered to be for supply and fix.

2. The material price in the original BOQ Rate to be as per PC Rate.

3. A uniform percentage of 3% shall be used for wastage if applicable.

4. The actual material price of the item to be within the PC Rate and cannot exceed same. Therefore, all PC Rate adjustment shall lead to only savings.

5. Minimum of three quotations to be obtained before finalizing the purchase price.

6. This is a varied work therefore a VO to be issued. The entire item of work to be removed from BOQ (with the note of VO.

No.) and the same to be included under Variations., with new (adjusted) Rate

7. The labour & Plant to be calculated based on the material component of 1.03 of PC Rate, if waste is applicable.

8. A lump sum per m2 for other materials such as grout,powder etc to be used.

9. The labour(L),Plant(P) to be obtained from the price analysis schedule and to remain constant during the subsequent adjustments.

10. The percentages of Overhead and Profit based from the Price Analysis schedule Or the contract to be used for adjustments.

The following Formula, with the only variable(purchase price) can be used.

m (constant) = PC Rate , Y(Constant) = 100-Over head-Profit

k (Constant) = Labour + Plant+ other materials M(Variable) = Accepted Invoice price

$$\frac{(1.03xm+labour+Plant + other\ materials)x100}{Y} = Original\ BOQ\ Rate = R$$

The Adjusted Rate shall be

$$\frac{1xM+k}{Y} \times 100 \quad or \quad \frac{1.03xM+k}{Y} \times 100, \text{ if there is wastage}$$

Example:

1. BOQ Rate 17/m2, PC Rate 15/=per m2,,Overhead 4%,Profit : 8% (From Price Analysis Schedule). Accepted Invoice price (M) = 10/=per m2

   $Y = 100-4-8 = 88$

   $K = \dfrac{17 \times 88}{100} -1.03 \times 15 = -0.49$

Check the Formula:

$$\dfrac{1.03 \times M + k}{Y} \quad x \quad 100 \text{ as there is wastage.}$$

$$\dfrac{1.03 \times 15 \quad -0.49}{88} \quad x\ 100 = 17.00$$

The BOQ Rate., with the given PC Rate.

The Adjusted (New) Rate shall be

$$\dfrac{1.03 \times 10 \quad -0.49}{88} \ x\ 100 = \textbf{11.15/m2}$$

Note: This Formula can be used whether the Constant K (Labour + Plant + other materials) is positive or negative.

However,it is more useful when the constant K is negative,where it is assumed that the other portion of this cost is priced elsewhere. The understanding is that the error is commited by the Contractor and hence has to bear the consequences.

3.5.5 AUDITING PROTOCOL—ANNEXURE FIVE
—RECOMMENDED PRACTICE

ADJUSTMENT FOR PRICE ESCALATIONS DURING CONTRACT PERIOD

It is common knowledge that prices of all items are increasing on a day by day basis .The effect of which is more apparent in the Construction industry as a project has a considerable duration from start to completion.This effects the pricing of projects as the Contractors are unable to guess to some degree of accuracy the possible escalation of prices during the construction period.

Most projects are traditionally fixed price Contracts as there was no such substantial escalations were found in the previous years,even if there were, the same were marginal and practically made not much financial effect to the Contractors.

We are now witnessing big boom in the construction industry and the trend is set to continue in the near future as more and more big projects are in the pipe line.This along with the international financial effects such as price rises in oil, raw materials,equipment etc.there is a necessity or even a necessary evil to introduce the price escalation clause in construction projects.The logic behind the introduction of such a Clause in Contracts is to strike a balance in the Contractor's risk for the escalation of prices.The absence of such a Mechanism leads to lot of speculations regarding the price increases by the Contractor.In the unlikely event of a decrease in prices or stabilation of prices,the Client is not benefited as the Contract price has already been fixed and there is no Contractual provision for the Client to share such benefits. Thus, the concept of best value for money is not achieved.

The Escalation Clauses are widely used in International Contracts in many countries for decades.There are various formulae in place and the same are revised on a regular basis. Therefore,the Escalation Clause in construction projects is a present day necessity.

Before the introduction of such a formula/e,the following to be considered.

a) It is virtually impossible to apply the escalation of prices on all items of construction.As a start,important materials,say **limited to 5-10 nos** such as steel, cement and cement products,cables,major pipes,aluminium,fuel etc.which are subject to wide price variations to be considered.

b) The above scnerio in the construction also invites lot of speculators with artificial price increases in addition to the actual.This requires a Government agency such as Central Bank to fix prices of such major items on a monthly basis based on market surveys,both national and international. Such Government Agency fixed prices to be considered for the Price escalation.

c) Method of operation of the Escalation is as follows:

Project size: Price escalation is applicable to projects say above US $ 5 million with duration exceeding say 180 days.

Items for Price Escalation: To be specified. Say Reinforcement Steel, Structural steel, concrete with grades, cables etc.

Provisional amount for Price escalation:	An Amount to be specified in the Contract for such adjustment. All escalations both positive and negative to be effected under such bill.
Base Rates:	These are the Contractual Rates. The Rates to be the Rates given by the Government Agency for the start month of Construction.(eg. Date of commencent is 15thMarch 2013, then the basic Rates are the ones applicable for March 2013)
Sequence of Work	Including the ordering of materials to be specified and agreed. The Client is not expected to pay the additional cost owing to Contractor delay in ordering or the completion of the project.
Price Escalation	The actual quantity of material(for which the escalation Clause is considered) used for the particular month to be considered.The Final Account for Price Escalation shall be the total all monthly/Interim adjustments.

3.6   TYPICAL WORK ACTIVITIES OF A TECHNICAL AUDITOR

Understand the types and Conditions of Contract

Analysis of Tender Reports both at Award and after completion

Plan suitable Auditing programmes

Review Tender and "As Built Drawings" in relations to additions and omissions.

Review Extension of Time to Contractors

Review Contractors' Claims

Review Consultants' Claims

Review New Rates—Reasons & values.

Review Variations –Reasons & values.

Truthfulness of the Engineer's Report on Schedule date of completion,in the event there is an extension to original completion date.

Review Statement at Completion in relation to the truthfulness of Quantities and Rates.

Review Final Statement. in relation to the truthfulness of Quantities and Rates.

Preparation of Audit Reports, highlighting short falls, the resulting savings,lessons learned and recommendations for future improvements.

3.7 ROLES OF THE ENGINEERING INSTITUTIONS

1. Make the Engineering Professionals aware of their Professional duty to the Society to give the best possible value for the Tax payers money,which are spent on Government construction projects.

2. Review and update Auditing Standards on a continuous basis.

3. Design and update suitable training programmes for Engineers in order to better serve the Construction Auditing Industry.

4. Review the Auditors reports and take up important issues with relevant authorities.

5. Conduct Seminars,workshops regarding Technical Auditing to increase the awareness among Engineers and the general public

6. Accrediation of Construction Auditors

7. Construction has become global now,so there is a necessity for all Professionals to have a universal system of Technical Auditing .It is the responsibility of the Cost Professional Institutions locally to convince their respective Auditor Generals regarding the importance of Technical Auditing and work with similar organisations globally to enforce such system.The necessary approaches to be made to the United Nations organisations,World Bank,World Trade

Organisations etc for the recognition and subsequent enforcement of the system on a global basis.

8. Do regular researches regarding the possible scope of construction and the availability of construction materials in order to maintain a balance between supply and demand in order to keep the market speculators at bay.Advice Governments regarding any short falls or problems that may be anticipated in the future and the corrective actions that are to be taken.

9. Keep in touch with production and distribution of new construction related materials which give much better value for money

10. Update the construction professional regarding new developments such as Green Building concept, new project management tools etc.

11. Preparation of Technical Auditing Courses for for Accrediation.A model is given below:

3.7.1 MODEL COURSE FOR TECHNICAL AUDITORS

Entry Requirements

Candidates should have successfully completed a recognized Degree course in one of the Engineering fields.

Civil Engineering
Mechanical Engineering
Electrical Engineering
Quantity Surveying
or any other Degree Course approved by the Auditor General

In addition they should have a minimum of Three years active on the field experience.

Candidates are expected to follow the undermentioned course of study and pass the relevant examination followed by an interview to be appointed as Accredited Technical Auditor.

Module One: **3 Hours of study**
Introduction to Technical Auditing
Types of duties and Code of Ethics for Auditors

Module Two: **4 Hours of study + 3 Hour Tutorial**
General understanding of types of Contracts and Conditions of Contract.
Tender Conditions

Module Three: **4 Hours + 3 Hour Tutorials**
Tender Analysis
Award of Contracts
Post-Contract procedures &Contract documentation

Module Four: **3 Hours + 2 Hour Tutorial**
Initial Auditing

Module Five: **3 Hours + 2 Hour Tutorial**
Intermediate Auditing

Module Six: **4 Hours + 3 Hour Tutorial**
Final Auditing

Module Seven: **1 Hour**
Audit Querries

Module Eight: **3 Hours + 2 Hour Tutorial**
Audit Reports

Module Nine: **4 Hours**
Construction Frauds including two case studies

Module Ten
An assignment consisting of 3,000 words (approx.) on the topic of
Technical Auditing for Construction Projects – 25% of marks shall
be allocated for this.

A Three hour written examination covering the above topics shall
be conducted and candidates are required to score above 50%.

The Auditor General to have sufficient cadre for Technical Auditors along with a promotional structure.The criteria for promotion to be examinations and other outstanding activities by the Technical Auditors.

MODEL TUTORIAL QUESTIONS

Students should read carefully the question before answering. The answers to be precise and to the point. The time requirement is around one hour per question and answers to be approximately 600 words.

Question No. 1.

In a Shopping complex project ,a Lump sum of Dhs. 1,200,000 was given for cabling. During the execution of the project the entire cabling was changed partly owing to the short coming of the Electrical design and changes to the location of the Chillers. The Contractor submitted a Claim of Dhs.2,100,000 as additional based on their original tender estimates. While analyzing the

Contractor's entitlement in detail, the following came to light.

1) There is a definite increase in the cabling scope.

2) The higher capacity cables (4C x300 Armored cable) were increased substantially(From 90 m in tender to 1,000m actual)

3) The Contractor's tender price breakdown for Cables(based on their lump sum of 1,200,000/=) was found to be very high, around 300% of the prevailing market prices at the time of tender.

Discuss how the Contractor's entitlement can be assessed , which is fair and reasonable under the above circumstances. Give your comments/reasons.

Question No. 2.

A major Shopping complex Re-furbishment was tendered as a lump sum Contract.The project was tendered with a Bill of Quantities as a uniform document for tendering purpose and the tenderers were given the option of clarifications of the same.However,the project was awarded as a lump sum contract.During the course of implementation, the Client wanted some changes to flooring and the same was instructed as a Variation. While evaluating the Contractor entitlements the following came to light.

1) There are errors in the measurement of this BOQ item in the tender documents.

2) The Contractor deleted the BOQ quantities,and included his quantities,with unit rates and arrived at the amount for this item of work.There are errors in the Contractor measurements.

3) There is no change in the tender and as Built drawings, meaning the total area involved is the same.

4) The Engineer found the contractor unit Rates as reasonable and the same were used as a base for the determination of new Rates in the variation.

The details are given below:

Tender BOQ Area	Qty.	Contractor Amended as	Correct Qty.	Actual done
Flooring in Zone 1	2,000 m2.	1,500 m2	2,200 m2	2,200 m2
Flooring in Zone 2	1,200 m2.	1,100 m2	1,600 m2	1,600 m2

Flooring in Zone 3	800 m2.	1,400 m2	1,800 m2	1,800 m2
Flooring in Zone 4	4,050 m2.	3,700 m2	4,200 m2	4,200 m2
Flooring in Zone 5	3,100 m2.	3,200 m2	3,800 m2	3,800 m2
Flooring in Zone 6	1,500 m2.	1,400 m2	1,800 m2	1,800 m2
Totals	**12,650 m2**	**12,300 m2**	**15,400 m2**	**15,400 m2**

Tender amount

Flooring in Zone 1	1,500 m2	@ 150/ (PC Rate 100/ m2)	2,200 m2
Flooring in Zone 2	1,100 m2.	@ 180/ (PC Rate 120/ m2)	198,000.00
Flooring in Zone 3	1,400 m2.	@ 180/ (PC Rate 120/ m2)	252,000.00
Flooring in Zone 4	3,700 m2.	@ 190/ (PC Rate 125/ m2)	703,000.00
Flooring in Zone 5	3,200 m2.	@ 175/ (PC Rate 110/ m2)	560,000.00
Flooring in Zone 6	1,400 m2.	@ 200/ (PC Rate 140/ m2)	280,000.00
Totals			**2,218,000.00**

Actual amount after Variation

Area	Qty	Revised Unit Rate	Amout
Flooring in Zone 1	2,200 m2.	@ 173/ (PC Rate 120/ m2)	380,600.00
Flooring in Zone 2	1,600 m2.	@ 203/ (PC Rate 140/ m2)	324,800.00
Flooring in Zone 3	1,800 m2.	@ 203/ (PC Rate 140/ m2)	365,400.00
Flooring in Zone 4	4,200 m2.	@ 219/ (PC Rate 150/ m2)	919,800.00
Flooring in Zone 5	3,800 m2.	@ 198/ (PC Rate 130/ m2)	752,400.00
Flooring in Zone 6	1,800 m2.	@ 223/ (PC Rate 160/ m2)	401,400.00
Totals			**3,144,400.00**

The Engineer has recommended an amount of 926,400/= (3,144,400-2,218,000) as Variation for the above changes.

Discuss whether the Engineer's recommendation for the Contractor's entitlement can be accepted or modification are required, assuming the revised unit Rates are acceptable and there is no change between the tender and "As Built Drawings" for the flooring.

Question No. 3.

In a Govt. park project a variation has resulted for the change of the Protection slab laid over the drainage pipes. The details are as follows:

The Protection slab overDrainage pipes was given as a slab of width **300 mm with a thickness of 50 mm.** An Item for this was allowed in the BOQ. and the same was priced as **26.00/m.**

During the execution of the Project, the Engineer changed the Protection slab as per Drainage Department standard details. In essence, the Protection slab was made to a width of 1500mm , 150 mm thick reinforced with 10 mm dia.Re-bar at 200 mm c/c in both directions.

The Contractor submitted a new Rate 390.00/m based on Pro Rata from the available BOQ Rate. The Contractor submitted his analysis of new Rate as follows: .

Approved BOQ Rate for
300mm wide X 50 mm thick pre—cast concrete slab laid over drainage pipe was 26.00/m

Volume of Concrete for 1.00 m length as per BOQ requirements

300 mm X 50 mm = 1.00 x0.30 x 0.050
= 0.015 m3

Volume of Concrete for 1.00 m length as per new requirements:

1500mmx150 thick = 1.00 x1.500 x0.15
=0.225 m3

Rate for the new concrete paving slab on Pro Rata basis is

26/0.015 x 0.225 = 390.00/m.

This new Rate was approved by the Engineer as reasonable and the same was forwarded to Client for formal approval.

Discuss whether

a) **Whether the method used in calculating the new Rate is appropriate or not?. Give reasons.**

b) **If not appropriate ,propose a new Rate with analysis.**

c) **Assume the following all in Rates for supply & fix: mass concrete 280/m3, Reinforced Concrete 300/m3, Re-bar 3/ kg., Formwork 30/m2,excavation 10/m3,disposal 12/m3.**

3.8   CONCLUSION

It has been proved beyond any doubt that frauds do occur in the Construction industry world wide.The Auditor General's report in 2001 in the United Kingdom speaks about 10% loss owing to Frauds in construction. If that is the case in a developed country with all kinds of checks and balance,it goes without saying that construction frauds are in much bigger propotions in developing countries.

The construction process in general is now much complex and involves many entities and traditional auditing by Accountants is somewhat outdated.

Further,inspite of the potential for construction, International investors are still reluctant and unwilling to come in construction in a big way as there is no confidence that their money is in safe hands and shall at the end of the day be given the best value for money.Even the present turbulent nature of the traditional investment options such as soverign bonds and shares have not tempted the investors to construction. Technical auditing will go a long way in installing that confidence in the construction industry.

The primary benefits from audit work is the effective resolution of the audit findings and implementation of the audit recommendations.strengthing controls and continuous improvement to generate increases in efficiency and profitability over the lifetime of the project and the organization. As the role of project control involves educating the project team,managing project cost,time,addressing risk and change and accurately reporting project status,the audit function is an essential part of the project controls.

Countries are spending lot of money on construction and for example the second biggest labour force(the biggest being agriculture) in India is in construction.With the volume of planned huge infrastructure projects,the chances are that in the next decade,construction shall employ the biggest labour force in India.

According to recent media reports,a plan for a development bank by BRICS countries is taking shape and it is assumed in the Financial circles that with the proposed financial resources it will rival the World Bank.Needless to say that Technical Auditing is very important as this bank is intended for infrastructure constructions,in countries where the Accountants are still in charge of construction Auditing.

A word of caution is required here.The planned bank is for construction and therefore is expected to be involved in huge construction.A situation may be created,where there can be shortage of building materials,again the same can be exploited by speculators.Therefore, it is important that the proposed bank should have industries which produce building materials or should invest in such industries for their expansion with agreement to supply materials for their projects at pre agreed reasonable rates for long periods.

There were instances during the construction boom in the Middle East,where steel prices went up around 300% during a short period. This coupled with increases in the prices of cement and copper made some contracting companies to go bankrupt as the Contracts in that part of the world do not have provision for escalation clauses in the conditions of contract.

Unfortunately,the effects are still being felt in the Middle East. Some Employers encashed the Performance Bond of the

Contractor,when the Contractor is unable to do the project owing to such high fluctuations in the material prices.This resulted in the banks to insist the full amount of Performance Bond to be deposited in the bank in order to obtain same.This has affected the Clients as they have to pay higher amounts for same and the Contractors are finding it difficult to obtain performance bonds and hence unable to tender.This is another additional loss to Client as the competition is reduced in addition to high cost of performance bond. A solution to this problem is an urgent need owing to the amount of constructions taking place in that part of the world,and the Conditions of Contract are such that the Performance bond is kept upto the issuance of the Defects Liability Certificate. The solution to this problem is to have insurance cover as the premium for such cover is very small or the conditions of performance bond to be amended to suit the present day requirements.There is a requirement for the Technical Auditor to see the problems faced by all parties in construction and suggest suitable solutions.

The good sign is that the recent Global Financial Crisis has heightened the importance of project cost management /controls. It is also noted that global acceptance of Technical Auditing of construction projects shall be greatly assisted through official recognition of the same by major international organisations such as the World Trade Organisation and the World Bank. It is of paramount importance for Professional Institution which have a stake and who are professionally responsible to give "Value for Money in construction to have proper systems to Technically Audit the Construction projects, by competent and accredited professionals. It is also important for such Institutions to do regular researches to study the volume of constructions world wide along with the supply of materials to see that there is a balance in order to mitigate shortages as speculators are very active to exploit such situations.The example of which was seen by the closure of certain contracting companies during the recent construction boom in the Middle East. The Institutions should advice governments about the problems if any in this regard with corrective actions that are required.

The next decade is expected to witness massive construction in the region of US $ 12 Trillion a year.China,India and other emerging economies are expected to make more than 50% of such construction,making them the status of developed countries. As stated before the construction sector in terms of cost & time is not that satisfactory at present and without Technical Auditing in one form or another,the situation shall be the same in the forseable future..These countries should have proper Technical Auditing Systems as a matter of urgency as going by the industry standards,the construction frauds which is around 10%,we are in for not million or billion but in excess of US $ One Trillion dollar issue. To put in context, the amount is equivalent to that of the combined GDPs of Saudi Arabia and South Africa.

A good beginning is made by the Chartered Institute of Building regarding the recognition of time overruns in the construction by launching the world first Time Management Contract for Complex Projects.After two years in the making on 23 rd April 2013,CIOB shall publish the same.

Speaking about the Contract,Keith Pickavance,a past president of CIOB and lead author of CPC 2013 said "This is a modern day contract designed for data age.It underlines and meets the need for a collaborative and competent approach to how risks are managed utilizing transparent systems of data.It can be used with or without,Building Information Modelling and has been drafted to work in any country and legal jurisdiction around the world:He further added "The causes and consequences of delay are the single most common reason for uncontrolled loss and cost escalation in complex building and Engineering projects,where the design is produced by the employer or Contractor.CIOB followed this up and launched a Project Time Management Certificate course for the intended practioners.

The need of the hour is something on the similar lines to be developed by the Cost Engineering Professional Institutions for cost overruns.

There is no uniformity in construction projects regarding the Advance payment and the procedure to be followed in the event of exclation of prices during the Contract period.There is a necessity to maintain some uniformity and the Author feels that the following shall bring some stability and attractive contract prices in the construction sector.

All construction projects to have a condition for the Employer to pay say 10% of the Contract price as Advance payment with good enforceable conditions included bank guarantee from the Contractor.

All construction projects with construction period more than 180 days to have provision for the adjustments for excalation of prices for some selected price vulnerable items during the Contract period.

The Auditor General is a highly respected position in any country. Generally given wide constitutional powers to enforce laws which are suitable for the Government expenditure.The position is so important that in China,the same is proposed by the prime minister and approved by the National People's Congress.It is time that Technical or Performance Auditing of construction projects is given its due place.

The Technical Auditing proposal given in this book is a model or some form of guidance for developing countries. The same to be developed and expanded on a continuous basis to achieve the desired result of "value for money" in construction.

As the Client pays for the auditing, some may argue as to why his faults to be considered. The author wants to be fair and that is why not only the Contractor faults but also Client/Engineer faults are dealt in this exercise. The understanding is that the Contractors never miss their entitlements and always fight for their rights. It is the author's intention that, while assessing the Contractor entitlements, the auditor should be fair and understand that all parties can make mistakes.

The Auditing exercise always looked upon badly or negatively.This exercise is only to be looked in the context of Check and balance as it is the rule of nature and the rules of creations that nothing is perfect and nobody is perfect.

It is a professional and moral requirement to see that the funds,particularly public funds which involves vast sum are spent wisely.

Since nothing is perfect, there is always room for improvements and as projects of similar in nature are done repeatedly,there is always a probability if not a strong possibility to make corrections next time around on similar projects.

The stake holders client, consultant, contractor should not worry about auditing as this is a requirement and not meant to find fault with anybody.Actually,contractors and consultants to be proud to be audited and found correct,thus giving value for the money spent.

Technical auditing is all the more important when it comes to situations,where everything is urgent-concept of what is given to-day was required yesterday.This state of affairs that creates among others Employs or forced to accommodate sub-standard consultants. Work commences before the finalization of design resulting in high provisional sums and variations.

There is another genuine contractor concern that this Technical Auditing can take additional time to receive their payments. Provision in the Contract to be made for the Contractor to submit the Statement at Completion as early as practicable and the finalization of Audit to be reached before the issuance of the Defects Liability Certificate. Provision can also be made for interim payments during the auditing period.

There is a school of thought that this is an additional expenditure (although as little as 0.10%of the project cost) to the Client as some form of auditing is already done. What is done as Auditing now is Financial and Technical Auditing is considered to be Performance Auditing. History has shown worldwide that the Client never loses in the event of Technical Auditing as Technical Auditing always brings Savings. This is a Win-Win situation for Clients.

As it is established that irregularities,frauds etc. are taking place and substantial amounts are lost in construction,it would be interesting to note that how much of savings are generated by the present internal and external auditing systems. It can be safely assumed that if the existing double auditing do not bring in atleast 0.5% savings (as the loss is estimated to be 10%) and a set of proposals for improvement on a regular basis,then the existing has to be changed or need a complete overhaul.

4

PROFILE OF THE AUTHOR

The Author is a Chartered Quantity Surveyor & Arbitrator with over 40 years of experience.

The author has a rare mixture of Technical Qualifications coupled with long years of experience in the field and in teaching is in an unique position to understand both the need of the Industry and the students.

E mail: almameer@hotmail.com
        almohdameer@yahoo.com
Tel + (94) 67 2277625
   + (94) 770615754
Website : www.almameer.com

PROFESSIONAL QUALIFICATIONS:

Professional Member of the Royal Institution of Chartered Surveyors, London **(MRICS)**

Associate Member of the Australian Institute of Quantity Surveyors **(AAIQS)**

Associate Member of the Chartered Institute of Arbitrators, London **(ACIArb)**

Successfully completed the Civil Engineering Course in 1969.

Elected as an Associate Member by the Construction Surveyors' Institute, London in 1980 (A.M.C.S.I.)

Elected as a Member by the Institute of Engineers & Technicians, London in 1981 (M.I.E.)

Successfully completed the Technical Teachers' Training Course conducted by the University of Moratuwa, Sri Lanka in 1976.

EXPERIENCE

Quantity Surveyor — **35 years**

Teaching — **5 years.**

Countries worked — **Sri Lanka, Oman, Bahrain, Saudi Arabia, U.A.E.**

Personal contributions.

Presentation of a research paper titled "Auditing of Construction Project Final Accounts at the Pacific Asia Quantity Surveyors Congress."

Submission of Proposals for Technical Auditing of Construction Project Final Accounts.

Submission of Proposals for the Introduction of Quantity Surveying in the Technical College curriculum in Oman.

Publication of an Article under the title "Construction as an option to international Investors" in Building Economics (AIQS—December 2011). Thus making the QSS to think that they can earn in excess of US$ 100,000/= a week. as Financial Advisors.

Involved in Teaching Quantity Surveying courses at Technical Colleges in Sri Lanka.

Wrote the following books:

"How to be successful at competitive Interviews"
incorporating his successful QS interview experiences
(ISBN-10:184903205x & ISBN-13: 978-1849032056.
Published by Schiel & Denver Publishing Limited 25.02.013)

How To Win And Manage Construction Projects
(ISBN: 9781481785242 & 9781481785259.
Published by Authorhouse UK)

Irregularities,Frauds and The Necessity of Technical
Auditing in Construction Industry
(ISBN: 9781481799751 ,978148179974-4 & 978148179976-8.
Published by AuthorHouse UK)